THINGS THAT GROSS ME OUT!

# POOP IS SO GROSS!

BY LYNDA ARNEZ

Gareth Stevens PUBLISHING

**Please visit our website, www.garethstevens.com. For a free color catalog of all our high-quality books, call toll free 1-800-542-2595 or fax 1-877-542-2596.**

**Cataloging-in-Publication Data**

Names: Arnéz, Lynda.
Title: Poop is so gross! / Lynda Arnéz.
Description: Buffalo, NY : Gareth Stevens Publishing, 2026. | Series: Things that gross me out! | Includes glossary and index.
Identifiers: ISBN 9781482471410 (pbk.) | ISBN 9781482471427 (library bound) | ISBN 9781482471434 (ebook)
Subjects: LCSH: Feces–Juvenile literature. | Excretion–Juvenile literature. | Human physiology–Juvenile literature.
Classification: LCC QP159.A76 2026 | DDC 612.3′6–dc23

Published in 2026 by
**Gareth Stevens Publishing**
2544 Clinton Street
Buffalo, NY 14224



Designer: Tanya Dellaccio Keeney
Editor: Kristen Nelson

Photo credits: Cover Burhan Alibas/Shutterstock.com; p. 5 (dog poop) xtotha/Shutterstock.com; p. 5 (deer poop) Cup Of Spring/Shutterstock.com; p. 5 (bird poop) foto_molka/Shutterstock.com; p. 5 (rat poop) SANDY.hrmn/Shutterstock.com; p. 7 alexandrabadashovaa/Shutterstock.com; p. 9 cunaplus/Shutterstock.com; p. 11 Yaoinlove/Shutterstock.com; p. 13 Bespaliy/Shutterstock.com; p. 15 New Africa/Shutterstock.com; p. 17 Red Stock/Shutterstock.com; p. 19 kwanchai.c/Shutterstock.com; p. 21 Anthony Robinson/Shutterstock.com.



Printed in the United States of America



CPSIA compliance information: Batch #CS26GS: For further information contact Gareth Stevens at 1-800-542-2595.

# CONTENTS



**Boldface** words appear in the glossary.

## Everybody Does It!

Every living thing does it—worms, whales, iguanas, and even you! It comes in different shapes, sizes, and colors. You don't want to make too little or too much of it. What is it? It's poop!

RAT POOP

BIRD POOP

DOG POOP

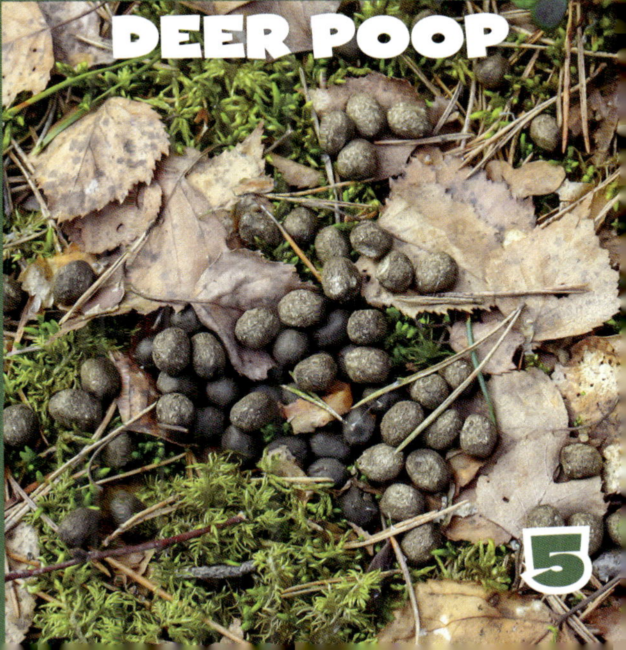
DEER POOP

## Making Poop

Poop is the body's waste. Making poop starts out with a living thing eating. The food moves through the digestive system. In people, this body system includes the mouth, **stomach**, and **intestines**. It breaks down food and takes in **nutrients** from it.

# THE DIGESTIVE SYSTEM

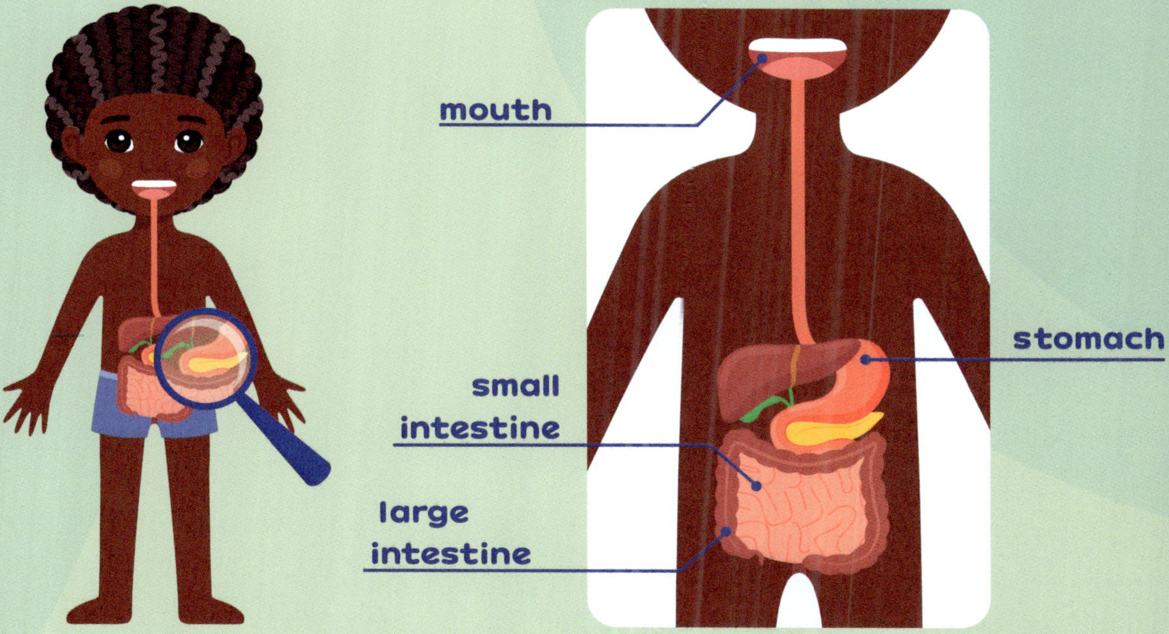

Poop forms in the large intestine. It is made up of what the body can't use or **absorb**. Poop is mostly water, **bacteria**, and **fiber**. We must poop to free up space in the body to digest more food!

## What a Smell!

Poop smells because of the bacteria in it. They give off gas as they help the body break down food. These bacteria are normal and needed in the body! Still, you shouldn't touch poop because of these bacteria. Wash your hands after you go!

## Colors

For people, poop should be shades of brown. The color comes from two kinds of matter in the body called bile and bilirubin. Some foods can change the color of your poop too. Poop should be somewhat smooth, or it can have some cracks in it.

# DOG POOP

## How Often?

Some people have to poop three times a day. Other people might poop every three days! This is all normal. However, if you ever notice changes in your habits, tell your parent or caregiver. Poop can tell us a lot about our health!

## Poop Problems

Diarrhea is a watery poop. You may have just one of these, or you might be stuck in the bathroom all day when you have diarrhea! It often goes away on its own after a few days. **Infections** can cause diarrhea. Eating food that has gone bad can too.

It's a big problem if you don't—or can't — poop! Being constipated means you aren't pooping enough. It may mean it hurts to poop, and your poop is quite hard. Getting enough exercise, drinking enough water, and eating more fiber often helps with constipation.

## Poop Eater?

Animals may *eat* poop! This could make them very sick. Sometimes, animals eat their own poop for a reason. Rabbits make a special kind of soft poop about once a day. They eat it to take in even more nutrients.

# GLOSSARY

**absorb:** To take in.

**bacteria:** Tiny living things that can only be seen with a microscope.

**fiber:** Part of food that the body can't digest and helps us poop.

**infection:** A sickness caused by germs.

**intestine:** A tube in the body that helps digest food.

**nutrient:** Something a living thing needs to grow and stay alive.

**stomach:** A part of the digestive system where food begins to break down.

# FOR MORE INFORMATION

## BOOKS

Daynes, Katie. *Where Does Poop Go?* London, UK: Usborne Publishing Limited, 2023.

Murray, Laura K. *Why Do We Need Poop?* North Mankato, MN: Capstone, 2024.

## WEBSITES

**How the Body Works**
*https://kidshealth.org/en/kids/center/htbw-main-page.html*
Find out more about human body systems here.

**Identify Poo**
*https://www.wildlifetrusts.org/wildlife/how-identify/identify-poo*
Check out animal poop and see how to identify it here!

**Publisher's note to educators and parents:** Our editors have carefully reviewed these websites to ensure that they are suitable for students. Many websites change frequently, however, and we cannot guarantee that a site's future contents will continue to meet our high standards of quality and educational value. Be advised that students should be closely supervised whenever they access the internet.

# INDEX